UNDER THE SEA ANIMALS

DOLPHINS

Amy Culliford

TABLE OF CONTENTS

A Pelican Book

Teaching Tips for Caregivers and Teachers:

Research shows that one of the best ways for students to learn a new topic is to read about it.

Before Reading

- Read the title and predict what the book will be about.
- Read the "Words to Know" and discuss the meaning of each word.
- Read the back cover to see what the book is about.

During Reading

- When a student gets to a word that is unknown, ask them to look at the rest of the sentence to find clues to help with the meaning of the unknown word.
- Motivate students with praise and encouragement.

After Reading

- Discuss the main idea of the book.
- Ask students to give one detail that they learned in the book.

Sight Words

a
all
and
eat
have
is
jump
like
move
play
some
their
this
to
use

Words to Know

blowholes

dolphin

fish

flippers

snouts

This is a **dolphin.**

dolphin

Dolphins like to jump and play!

All dolphins eat **fish**.

fish

Dolphins use their **flippers** to move.

flipper

Some dolphins have **snouts**.

snout

All dolphins have **blowholes**.

blowhole

Index

Written by: Amy Culliford
Design by: Under the Oaks Media
Series Development: James Earley
Editor: Kim Thompson

Photos: lauren marie baer: cover; dinerstein: p. 4-5; nikkytok: p. 7; Chanonry: p. 9; Ekaterina Kuzmenkova: p. 11; Iordache Elena G: p. 12-13; Sokolov Alexey: p. 15

Library of Congress PCN Data
Dolphins / Amy Culliford
Under the Sea Animals
ISBN 978-1-63897-071-2 (hard cover)
ISBN 978-1-63897-157-3 (paperback)
ISBN 978-1-63897-243-3 (EPUB)
ISBN 978-1-63897-329-4 (eBook)
Library of Congress Control Number: 2021945207

Printed in the United States of America.

Seahorse Publishing Company
www.seahorsepub.com

Published in the United States
Seahorse Publishing
PO Box 771325
Coral Springs, FL 33077